The Young Geographer Investigates

Islands

Terry Jennings

Oxford University Press

Oxford University Press, Walton Street, Oxford OX2 6DP

Oxford   New York   Toronto
Delhi   Bombay   Calcutta   Madras   Karachi
Petaling Jaya   Singapore   Hong Kong   Tokyo
Nairobi   Dar es Salaam   Cape Town
Melbourne   Auckland

and associated companies in
Berlin   Ibadan

Oxford is a trade mark of Oxford University Press

ISBN 0 19 917085 1 (limp non-net)
ISBN 0 19 917091 6 (cased, net)

© Terry Jennings 1988

First published 1988
Reprinted 1991

Typeset in Great Britain by
Tradespools Ltd., Frome, Somerset
Printed in Hong Kong

Acknowledgements

The publisher would like to thank the following for permission to reproduce photographs:

Aerofilms p.12 left; Bryan and Cherry Alexander p.15 bottom, p.37 bottom right;
Ardea p.21 bottom; Ashmolean Museum p.46 bottom; Aspect Picture Library p.4
top, p.8 top, p.12 right, p.31 centre right, p.41 left; Peter Baker p.30; Anne Bolt
p.38 bottom right; Bridgeman Art Library p.14; Camera Press p.32 top and
bottom left, p.35 top; Bruce Coleman COVER, p.7 bottom left, p.13 left, p.27,
p.40 left; Robert Estall p.15 left; John Frost p.30 right; Sally and Richard
Greenhill p.33 top and bottom, p.37 top; Robert Harding p.13 bottom, p.15
centre, p.28 bottom right, p.41 bottom right; John Hillelson Agency p.26, p.28
top, p.32 bottom right, p.36; Eric and David Hosking p.31 bottom right; Image
Bank p.4 bottom, p.6 top, p.7 top, p.11, p.31 left; Impact Photos p.16 top, p.41
top right; Terry Jennings p.21 top, p.23, p.40 top right; Natural Science Photos
p.7 bottom right, p.10 bottom, p.34 bottom, p.35 bottom; Photo Library of
Australia p.10 top; G.R. Roberts p.13 top, p.39 bottom; Rapho p.16 bottom left
and bottom right, p.17 bottom; Rex Features p.35 centre; Shostal Associates p.29
top and bottom, p.38 bottom left; Spectrum Colour Library p.39 top; ZEFA p.4
centre, p.17 top, p.34 top, p.38 top.

Illustrations are by Ben Manchipp, Ed McLachlan, Mike Saunders, Techniques.

Contents

Islands

Fair Isle, Scotland

An island is a piece of land with water all around it. Islands are completely separated from continents by the sea. There are also small islands in some lakes and rivers.

The world's largest island is Greenland. Other large islands include Papua New Guinea, Borneo, Baffin Island and Madagascar. Many more islands are so tiny they are difficult to see from an aircraft.

The number of islands changes from time to time. Small islands may be worn away by the waves. Eventually they disappear under the sea. Sometimes, as we shall see later in this book, new islands are formed. Volcanoes under the sea may form new islands. So may tiny animals called corals.

A lake island in Sri Lanka

Islands on the River Nile between Abu-Simbel and Assuan

4

Kinds of islands

There are two main kinds of islands. These are oceanic islands and continental islands.

Oceanic islands lie far out to sea. Ascension Island and St. Helena in the Atlantic Ocean are oceanic islands. There are also many oceanic islands in the Pacific Ocean. They include the Hawaiian Islands and the Galapagos Islands. Some of these oceanic islands are the tops of high mountains rising from the bottom of the sea. Others, such as Iceland and Tahiti were built up by volcanoes under the sea. These islands which are undersea mountains or volcanoes are often rocky with steep cliffs. In warmer seas, many oceanic islands are made of coral.

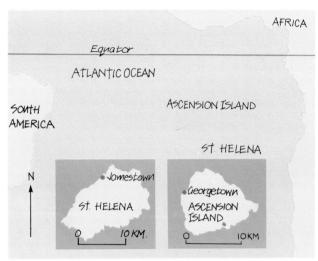

Oceanic islands usually have few kinds of plants and animals living on them. The main kinds of animals are birds and insects. These have reached the islands by flying to them. The plants on oceanic islands are mostly those whose seeds have been carried there by the ocean currents.

The tops of undersea mountains and volcanoes may appear as islands

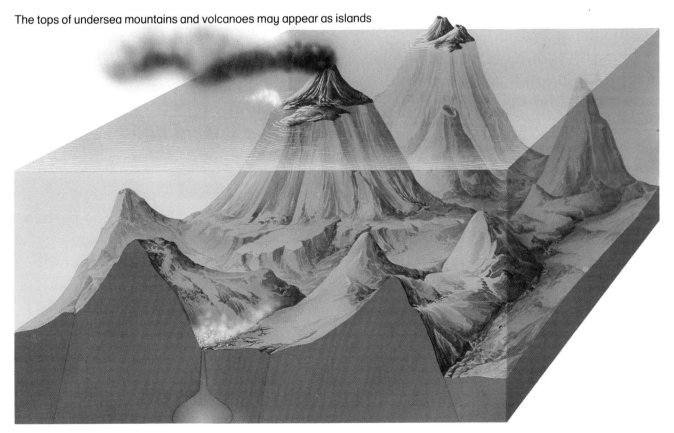

Continental islands

Continental islands off the coast of Oregon, U.S.A.

Continental islands always lie close to the mainland. Many of them were once joined to the mainland. They became islands because the sea level changed or because pieces of the Earth's crust moved.

Thousands of years ago, for instance, the world was much colder than it is today. The world was in what is called the Ice Age. During the Ice Age many of the seas were frozen. A lot of the land was covered by ice and snow. The sea level was much lower than it is today. At that time, with the shallower seas, the British Isles were joined to the continent of Europe.

When the Ice Age ended, the ice and snow melted. The sea level rose because of all the extra water. And the land between Great Britain and Europe was flooded, forming the North Sea and English Channel.

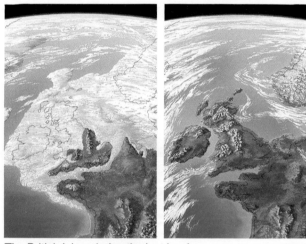

The British Isles, during the last Ice Age

Later Ireland was separated from Great Britain. In this way the British Isles became continental islands.

Other continental islands include Japan, Hong Kong and Trinidad. In places the land sank into the sea. Only the tops of mountains were left standing up above the water. Many of the islands off the west coast of Scotland and off the coast of Canada were formed in this way.

Coral islands

A coral island in the southern Pacific Ocean

Coral islands are a type of oceanic island. They are mainly made up of the skeletons of tiny animals. Coral animals live only in warm seas. These animals are related to sea anemones.

A single coral animal is called a polyp. Each coral polyp has a hard chalky skeleton. When the living polyps die, their hard skeletons are left. More polyps grow on top of the hard skeletons. And so the coral grows higher and higher.

Coral grows in many colours. It may be red, pink, yellow or white. As the coral grows it forms beautiful shapes. Eventually the coral may form a low island rising only a few metres above the sea. Or it may form a ridge in the sea called a reef.

On the sea-bed there are many mountains and volcanoes that are not quite high enough to stick up above the water. Sometimes coral grows up on one of these. Eventually a coral island is formed.

Fire coral growing on a sea fan

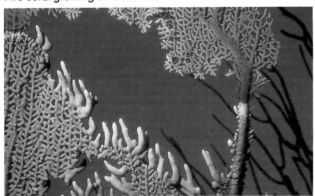

Brain coral in the sea off St Lucia, in the West Indies

Coral reefs

A coral island atoll

The coral polyps can live only in shallow water. Because of this they often grow around the shores of an island. When this happens the coral forms what is called a fringing reef.

Sometimes the coral builds up a little way out from the shore. This kind of reef is called a barrier reef.

The water between the reef and the island is calm and sheltered. It forms a lagoon. The largest barrier reef in the world is the Great Barrier Reef. It lies off the east coast of Australia. The Great Barrier Reef is more than 2000 kilometres long and contains many coral islands.

Sometimes, as we have seen, beneath the sea there are volcanoes. The coral may build up in the shallow water around one of these. Eventually a reef is formed. Later the volcano may sink below the sea. Only a circle or horseshoe-shaped reef is left with a lagoon in the middle. Such an island is called an atoll. Aldabra in the Indian Ocean is an atoll. Atolls are found in all tropical seas. But they are most common in the Pacific and Indian Oceans. There are over 300 atolls in the Pacific Ocean alone.

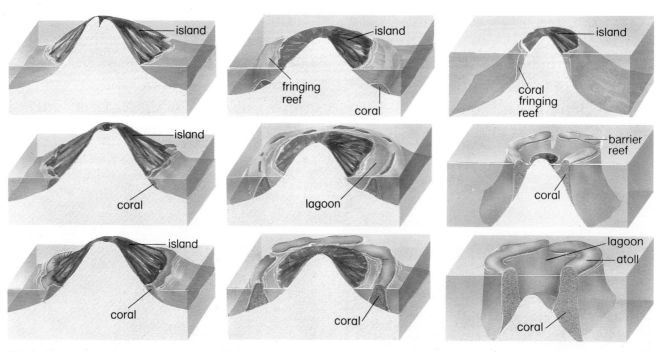

The formation of a fringing reef, a barrier reef and an atoll

Volcanic islands

The formation of a volcanic island

Many oceanic islands began life as volcanoes. One day a volcano on the sea-bed suddenly exploded or erupted. Molten rock or lava flowed out from cracks in the sea-bed. The lava piled up in the water and solidified. Eventually this new

Krakatoa – a volcanic island in Indonesia

volcanic rock rose above the surface of the sea. A new island had formed.

The Hawaiian Islands were formed like this. They are the tips of undersea volcanoes. Tristan da Cunha in the South Atlantic is an old volcano. So are the Galapagos Islands, the Azores, the Canary Islands and many more.

At first nothing can live on a volcanic island. But as the lava cools, plants and animals can live on it. Islands are constantly being worn away by the sea. If they are small they may last only a few years. But large volcanic islands can eventually have many plants and animals living on them. This often takes thousands of years, though.

New islands

The 'Twelve Apostles' stacks, Port Campbell, Australia

Ocean waves may help to form new islands. Waves crashing against cliffs, break or loosen pieces of rock. Gradually the cliffs wear away. Sometimes the sea wears away the softer parts of the rocks. It leaves behind the harder rocks. These may form small rocky islands called stacks. Some stacks are shown in the picture above.

Not only can the sea form new islands, so can certain plants. One plant which can do this is the mangrove tree. There are many different kinds of mangrove trees growing in the tropics. They all grow where the water is shallow. Mangroves are able to grow in the water because they have special breathing roots.

Mangroves, like corals, are sometimes island builders. In sheltered seas their roots trap mud and floating pieces of plants. Gradually the level of the mud is raised above the water. Then new dry land is formed. The mangroves may trap enough mud and debris to form a new island.

Mud islands are formed around the roots of the mangrove

Disappearing islands

A winter storm on the Oregon coast, U.S.A.

Waves crash against islands. Pieces may break off the rocks. Gradually the islands get smaller and smaller. After a long time, if the island is very small, it may be completely worn away by the waves.

But sometimes islands disappear quicker than this. The Aurores Islands were first discovered in the 1760s. They were situated in the South Atlantic to the south-east of the Falkland Islands. Men from a Spanish ship later mapped the islands. But a few years afterwards, the captain of another ship decided to visit the islands. When he got to the spot the islands were not there. Only waves could be seen where the islands had stood. Eventually the Aurores Islands were removed from the map.

A few years ago a new island was discovered off the coast of Sicily. It was given the name Graham Island. A short time after Graham Island had been mapped, it too disappeared.

Perhaps these islands were destroyed by a volcano. Possibly an earthquake shattered them. Whatever happened, these islands, and others, have disappeared never to be seen again.

Clusters of islands

The Scilly Isles

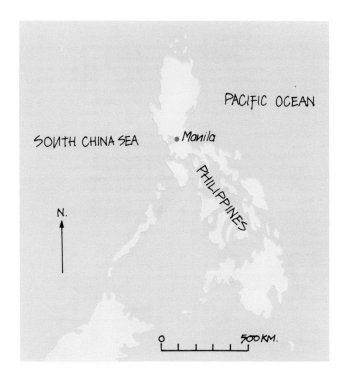

Some oceanic islands are thousands of kilometres from the nearest land. Often, however, islands are found in groups or clusters. A group of islands is called an archipelago.

The largest archipelago in the world is in the Pacific Ocean. It is a group of more than 13 000 islands. These form Indonesia. Some of the Indonesian islands are volcanic, some are made of coral.

There are many smaller archipelagos. The Galapagos Islands form an archipelago of islands in the Pacific Ocean. The Seychelles is an archipelago in the Indian Ocean. In the English Channel, between England and France, are the Channel Islands. These form another small archipelago. The largest islands of the Channel Islands archipelago are the well known Jersey, Guernsey, Alderney and Sark.

The Philippines, Canaries, Bahamas and Japan are just a few more archipelagos.

The Christmas Islands archipelago in the Indian Ocean

How do animals and plants get to islands?

After a new island has formed, it is slowly colonized by plants and animals. Birds and bats are able to fly to new islands. Seals can swim there. Insects, spiders and plant seeds may be blown by the wind. Dandelion seeds were blown by the wind from Australia to New Zealand. From there the seeds of the dandelion were carried to some of the islands of Polynesia.

Birds may carry snails and insect eggs in the mud on their feet and feathers. The mud may also contain the seeds of plants. Some plant seeds can survive long journeys in salt water. They include the seeds of coconuts and mangroves.

Motutara Island stack, New Zealand, with its gannet colony

Washed up on a beach, this coconut is now germinating

Most of the animals carried to an island probably die. However seabirds and seals can use new islands for breeding and resting. The other animals can only live there if plants are growing on the island. The plants can then provide the animals with food.

Grey seals on the Farne Islands

Occasionally huge clumps of plants are torn from the mainland. These are swept along by the ocean currents. Small animals may drift along on these clumps.

Islands and evolution

HMS *Beagle* in which Darwin sailed around the world, 1831–36, in what is now the Beagle Channel in southern Australia

Animals and plants on continental islands are very similar to those on the nearby mainland. But animals and plants on oceanic islands are often completely different. They may occur nowhere else in the world.

These rare plants and animals have always been of great interest to scientists. In the 1830s the British scientist Charles Darwin made a voyage around the world. On this voyage he visited the Galapagos Islands in the Pacific Ocean. He studied the rare plants and animals there. One of the things Darwin noticed was that each of the Galapagos Islands had different kinds of birds on them. These birds were different although obviously related to each other.

After his voyage, Charles Darwin explained how different kinds, or species, of animals and plants arise. When it arrives on an island a species may slowly change. Over thousands of years it may change into a new species that exists nowhere else. These changes are called evolution.

Varied bill evolution among Galapagos finches

Coconut palms

Many of the ripe coconuts are split open and allowed to dry. The white flesh is scooped out and dried still further. It is then called copra. Coconut oil can be pressed from the copra. This oil is used to make margarine, soap, cosmetics, detergents and many other products. The part of the copra that is left is fed to cattle. Dried and shredded coconut flesh (dessicated coconut) is also used in sweets and cakes.

Coconut palms and (inset) a fresh coconut cut in half

Right: splitting coconuts before drying them in the sun

Coconut palm tree leaves make a good roof

The coconut palm is very common on tropical islands. On the tree the hard brown coconut is enclosed in a thick layer of fibres. With these fibres around it, the fruit can float away if it falls in the sea. It was ocean currents as well as people which spread the coconut palm to so many countries.

The palm tree is one of the most useful tropical trees. Its trunk provides timber for building huts. The leaves are used to thatch roofs and to make hats, baskets and mats. The fibres around the nut are used to make heavy doormats and machine belts. Even the flowers are used to make a drink called toddy.

Sugar cane

Harvesting sugar cane by machine in Cuba

Most sugar is made from sugar cane, a giant grass. This plant is grown on many tropical islands, including the West Indies, Hawaii, Jamaica, Cuba and Mauritius. Usually the sugar cane is grown on large farms called plantations. Sugar cane grows best in rich moist soil. It also needs a lot of hot, sunny weather. Some kinds of sugar can grow to be over 6 metres high. The stems may be 5 centimetres in diameter at their bases.

The sugar cane is cut by hand in many places. But machines are being used more and more. The cut cane is quickly taken to the mill. There it is chopped into short pieces. The pieces of cane are crushed by huge rollers to squeeze out the juice. Finally the juice is boiled until it forms sugar crystals.

Cut sugar cane

Newly produced sugar

Fishing for food

Many sea fish are good to eat. And living surrounded by water, island people often earn their living by fishing.

One kind of fishing boat is called a drifter. A drifter uses long nets with corks or floats on them. The net hangs down like a curtain just below the surface of the sea. The boat and net drift along with the wind and tides. Fish swim into the net and get caught in it. Somewhat similar is purse-seining. In this, the fishing boat surrounds a shoal of fish with a curtain of netting.

A trawler is another kind of fishing boat. Trawlers can stay at sea for a long time. A trawler uses a net like a big string bag. The net is pulled along until it is full of fish. The fish are frozen to keep them fresh until the trawler gets back to port.

Lobsters, crabs and crayfish also live in the sea. Fishermen catch these shellfish in special pots or traps. The fishermen put some pieces of fish in

A trawler and (inset) fresh crabs

the pot and then lower it to the bottom of the sea. If a shellfish goes into the pot after a piece of fish, it cannot get out again.

A drifter, trawler, and purse-seiner

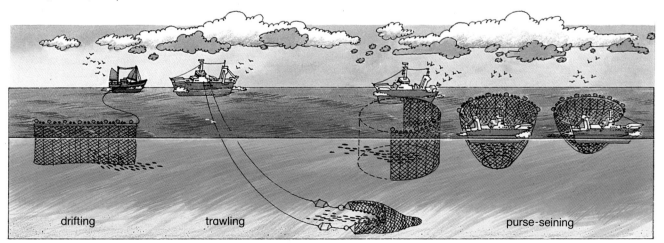

drifting   trawling   purse-seining

17

Do you remember?

1 What is an island?

2 What is the world's largest island?

3 Why does the number of islands change?

4 Where are oceanic islands found?

5 Name three oceanic islands.

6 Where are continental islands found?

7 What was the world like during the Ice Age?

8 Why did the sea level change after the Ice Age?

9 Name three continental islands.

10 What is a single coral animal called?

11 Where do coral animals live?

12 What is an atoll?

13 What is a lagoon?

14 How may a volcano under the sea form a new island?

15 How are sea stacks formed?

16 Why are mangrove trees able to grow in water?

17 What is an archipelago?

18 Name three archipelagos.

19 How may some plant seeds get to a new island?

20 How do fishermen catch lobsters, crabs and crayfish?

21 What is it called when a plant or animal species gradually changes into a new species?

22 What encloses the hard brown coconut when it is growing on the palm tree?

23 How is the coconut palm spread to new countries?

24 What are coconut palm trees used for?

25 What is copra and what can be obtained from it?

26 What kind of plant is sugar cane?

27 How is sugar obtained from sugar cane?

28 What kind of fishing net does a drifter have?

29 How does a trawler catch fish?

30 How may birds and insects get to a new island?

Things to do

1 Make a model continental island Choose a continental island which is very close to the mainland. The Isle of Wight, Newfoundland, Sicily, Anglesey, Sri Lanka or Jersey are just a few suitable ones.

Make a model of your chosen island and part of the nearby mainland. Make your model on a large sheet of card, plywood or hardboard.

You could, if you wished, use pieces of wire netting as the foundation for your island and the mainland. Crumple the wire netting into the shape you want.

(Careful — the wire may be sharp!).

Cut newspaper into strips about 2 centimetres wide. Mix a small bowlful of thin cold-water glue or wallpaper paste. Wet strips of the newspaper with the glue or paste. Cover the wire netting with the strips. See that all the wire netting is covered with several layers of newspaper.

Leave your model on one side. When your model has dried out completely, paint it. Paint the sea blue and the land green or brown.

Make tiny buildings from card. Stand clusters of these on your island and the mainland to represent the main towns.

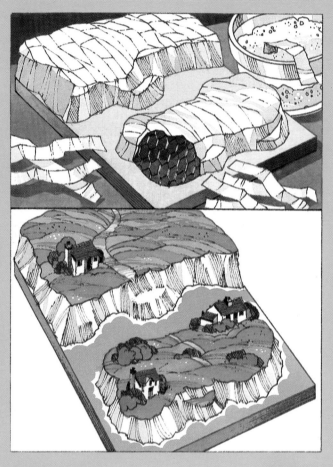

2 Make paper palm trees Paper palm trees can be made from rolled up newspaper or drawing paper. Stick the ends with glue or Sellotape. Carefully cut along the roll of paper at one end so that the strips form the palm tree leaves.

Colour the trunk of your tree brown and the leaves green. Use paints or felt-tipped pens for this. Fix the base of your tree trunk in a lump of plasticine so that it will stand up on its own.

3 Stories about islands There are a number of exciting stories about islands. Three of them are *Robinson Crusoe* by Daniel Defoe, *Coral Island* by R. M. Ballantyne and *Treasure Island* by R. L. Stevenson.

Write your own story about an island. Draw some large pictures to illustrate it.

4 A map of treasure island Have you ever seen pictures in story books of maps showing where buried treasure is to be found? Make your own map of treasure island on white paper. Give clues as to where to find the treasure. Tear around the outside of the map to make it look frayed and old. Usually old maps are a faded brown colour. You can make your map look very old by dipping it into cold tea and then leaving it to dry.

5 How rocks break down As we saw on page 11, the waves gradually wear away rocky islands.

You can see how water breaks up rocks if you do this simple experiment.

Break some chalk into small pieces (blackboard chalk will do). Chalk is a soft rock and will break more easily. It would take a long time for this experiment to work if you used harder rocks.

Half fill a bottle with water. Put some of the pieces of chalk in the water. Put the stopper on the bottle and shake the bottle hard for as long as you can.

Then look at the pieces of chalk. How do they differ from pieces that have not been shaken with water? What can you see at the bottom of the bottle?

You could also try this experiment with small pieces of brick. Do you get the same results?

Can you understand now how pieces of rock broken from cliffs are worn down to boulders, then pebbles, and eventually to the sand which finishes up on the beach?

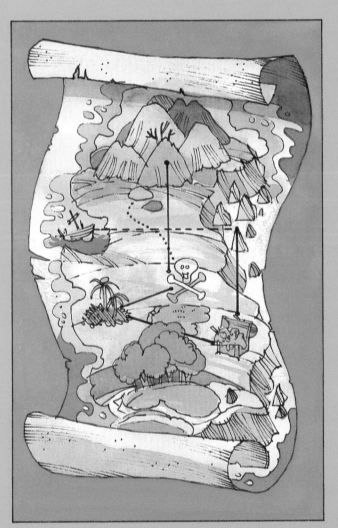

6 The work of the sea As we saw on page 11, the coasts of islands are gradually worn away by the sea.

If you visit the seaside, look for signs of damage done by the sea. Look at the cliffs and promenades, and any holes or cavities worn in the rocks or beach. Write down what you notice.

Make cliffs from sand or sand castles near the water's edge. Watch to see how the waves break them down.

7 Make a big picture of a fishing drifter or trawler Ask your friends to help you to make a big picture of a fishing boat. Use paints, scraps of cloth, tissue paper and sticky paper. Do not forget to show the nets and some of the fish the boat has caught.

8 A model fishing boat Make a model of either a fishing drifter or a trawler. Use cardboard boxes, tubes, drinking straws, empty plastic bottles and any other scrap materials you can find. Paint your model.

9 Coconuts for the birds Rather strangely, although coconuts grow in tropical places, birds in other parts of the world like to eat the white coconut flesh.

Blue tits may visit your coconut

Cliff fall in the Algarve, Portugal

Cut a coconut in half (Careful!) having first drained off the milky liquid inside. Then make a hole through the base of each half, and hang or nail each one upside down or sideways on the branch of a tree or on a stick. If you fix them any other way up, water will collect in the cup and the flesh of the nut will go mouldy. Watch to see which kinds of birds come to feed on the coconut flesh.

Do not, however, hang coconut in your garden during the spring or summer, since young birds cannot digest it. And **never** feed dried, shredded coconut to birds, young or old. This is really dangerous because it can swell up inside a bird and kill it.

When your coconut shell is empty, it can still be used as a bird-feeder. Mix up some bread or cake crumbs, seeds or cooked potato with warm fat. Fill the inside of the coconut shell with this mixture. The birds can then come to eat it.

10 The coconut's journey Pretend that you are a coconut growing on a palm tree on the shore of a tropical island. You fall in the sea when you are ripe, and are carried away by the ocean currents. Write a story describing what happens to you.

11 Planting seed travellers Many plants arrive on new islands by accident. You can discover some of the ways in which seeds can travel if you do this experiment.

You will need some shallow dishes such as old saucers or coffee jar lids, some blotting paper, paper towels or tissues, and some 'cling-film'.

After you have been for a walk across a park or in the garden or country on a wet day, carefully scrape the mud from your shoes. Line the saucers or coffee jar lids with several layers of wet blotting paper, paper towels or tissues. Sprinkle the mud from your shoes on to the wet paper, and then wet the mud. Cover the dishes with 'cling film'. Keep them in a warm place.

Do any seeds germinate from the mud? Can you transplant any of the seedlings into pots of soil to see what kinds of plants they grow into?

If you can obtain permission, get some of the mud from under the wheel arches of a car or under the mudguards of a bicycle. Are there any seeds in this mud?

12 Island-hopping around the world Imagine you are going to make a journey right round the world. You are going to travel from one island to another. Look at a map and decide which islands you will visit. Try to work out roughly how far it is between each pair of islands. What is the total length of your journey?

Write a story about your journey and the adventures you have.

13 Collect island stamps Many islands have very attractive postage stamps. Choose an island and collect as many of its stamps as you can. Display your stamps in an album or on a wall chart. Write a sentence or two about each of the stamps, saying what the picture on it tells you about the island.

Things to find out

1 Here is a map quiz. An atlas will help you to answer these questions about the map below:
(a) The map has four islands marked with their initials. What are their names?
(b) What is the name of the large mainland country which is shaped like a boot?
(c) What is the name of the city marked on the mainland country?
(d) What is the name of the mountain on the island nearest to the mainland? What kind of mountain is it?
(e) What is the name of the sea which surrounds all four islands?

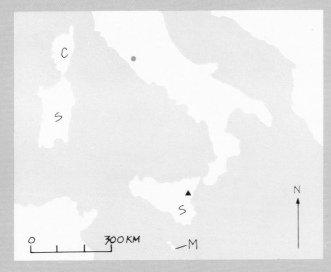

2 Islands usually have different climates from places in the middle of continents even when they are the same distance from the Equator. For example, Edinburgh which is on the island of Great Britain is the same distance from the Equator as Moscow which is far inland in the Soviet Union. But Edinburgh and Moscow have totally different climates. Find out why this is.

3 Here are the names of some famous volcanoes: Mont Pelée, Krakatoa, Mauna Loa, Fujiyama, Mount Etna, Mayon, Mount Erebus, Ruapehu. They are all to be found on islands. Find out the name of the island or island group to which each volcano belongs.

4 What are the islands people most often visit during their holidays? Look at travel agents' advertisements and brochures to find out. What are the attractions of these places to holiday-makers?

5 Find these islands in an atlas: St. Helena, Tasmania, Fiji, Victoria Island, Taiwan, Tierra del Fuego. Say which of these islands you think are oceanic islands and which are continental islands.

6 Find out how pollution of the sea affects coral and coral islands. What other things threaten the living coral?

7 The sand around the shores of islands and continents can be many different colours. Find out why different places have different coloured sands.

8 Sand dunes occur around the coasts of many islands. Find out how sand dunes are formed and what happens to them.

9 Waves and tides affect the lives of fishermen, sailors and holiday-makers on islands. Find out how waves and tides are formed.

23

Islands map

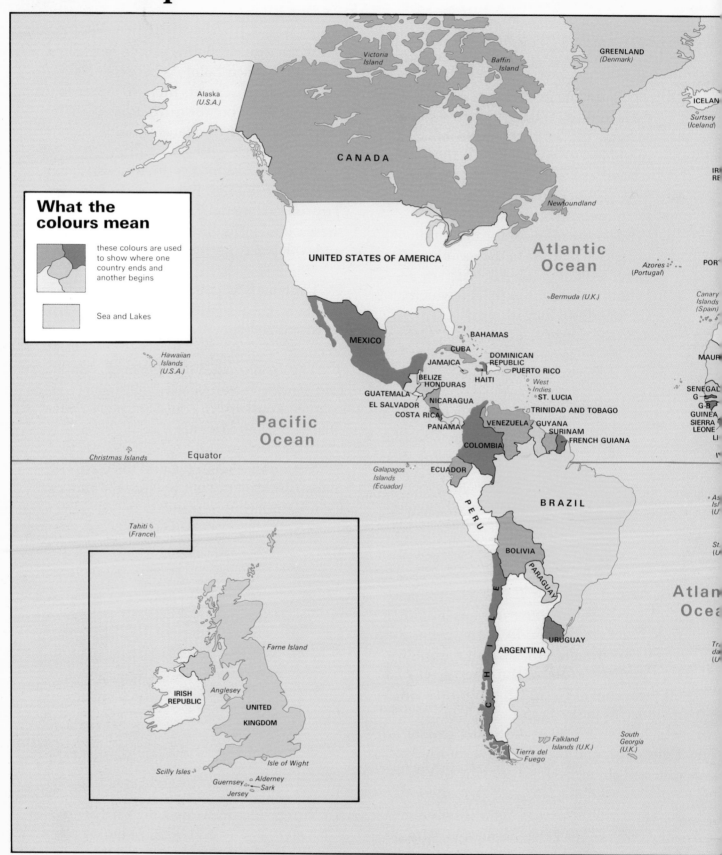

What the colours mean

these colours are used to show where one country ends and another begins

Sea and Lakes

Alaska
(U.S.A.)

Victoria
Island

Baffin
Island

GREENLAND
(Denmark)

ICELAN

Surtsey
(Iceland)

CANADA

Newfoundland

Atlantic
Ocean

IR
RE

Azores
(Portugal)

POR

Bermuda (U.K.)

UNITED STATES OF AMERICA

Hawaiian
Islands
(U.S.A.)

Canary
Islands
(Spain)

MAUR

MEXICO

BAHAMAS

CUBA

DOMINICAN
REPUBLIC

JAMAICA

PUERTO RICO

SENEGAL
G-

BELIZE
HONDURAS

HAITI

West
Indies

G-B

GUATEMALA

ST. LUCIA

GUINEA

EL SALVADOR

NICARAGUA

SIERRA
LEONE

COSTA RICA

TRINIDAD AND TOBAGO

LI

Pacific
Ocean

PANAMA

VENEZUELA

GUYANA

SURINAM

I

Christmas Islands

Equator

COLOMBIA

FRENCH GUIANA

Galapagos
Islands
(Ecuador)

ECUADOR

As
Isl
(U

P
E
R
U

B R A Z I L

Tahiti
(France)

BOLIVIA

Atlan
Ocea

Farne Island

PARAGUAY

St
(U

C
H
I
L
E

URUGUAY

Tr
da
(U

IRISH
REPUBLIC

Anglesey

ARGENTINA

UNITED

KINGDOM

Falkland
Islands (U.K.)

South
Georgia
(U.K.)

Scilly Isles

Isle of Wight

Tierra del
Fuego

Guernsey

Alderney
Sark

Jersey

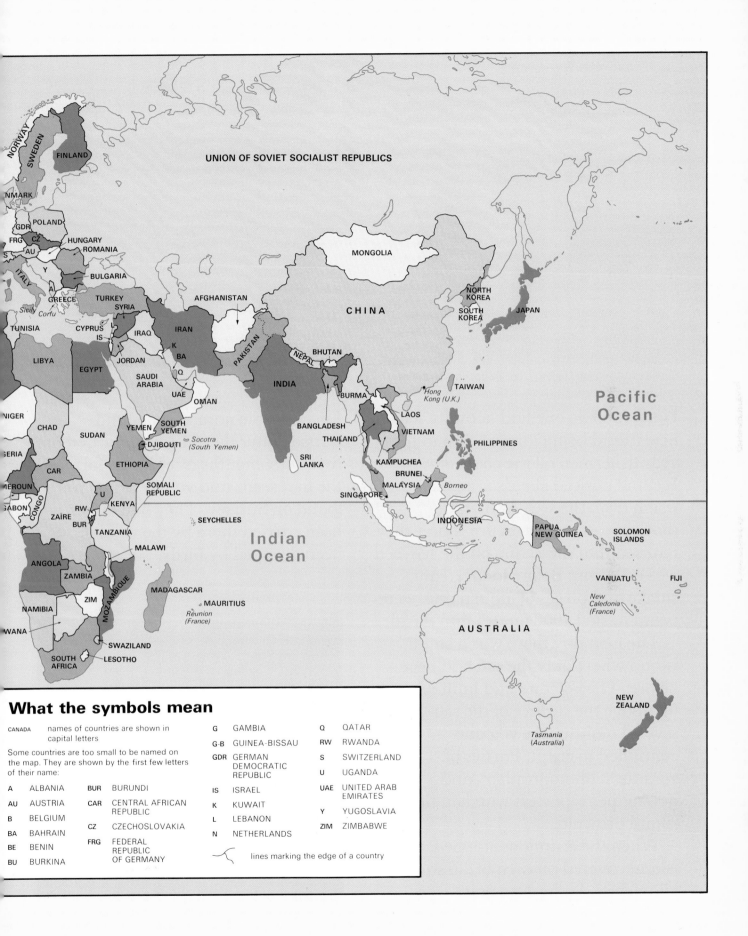

What the symbols mean

CANADA names of countries are shown in capital letters

Some countries are too small to be named on the map. They are shown by the first few letters of their name:

A	ALBANIA	BUR	BURUNDI
AU	AUSTRIA	CAR	CENTRAL AFRICAN REPUBLIC
B	BELGIUM	CZ	CZECHOSLOVAKIA
BA	BAHRAIN	FRG	FEDERAL REPUBLIC OF GERMANY
BE	BENIN		
BU	BURKINA		

G	GAMBIA	Q	QATAR
G-B	GUINEA-BISSAU	RW	RWANDA
GDR	GERMAN DEMOCRATIC REPUBLIC	S	SWITZERLAND
IS	ISRAEL	U	UGANDA
K	KUWAIT	UAE	UNITED ARAB EMIRATES
L	LEBANON	Y	YUGOSLAVIA
N	NETHERLANDS	ZIM	ZIMBABWE

lines marking the edge of a country

Surtsey

Surtsey erupting (and below) plant life appears on the new island after a few months

Until 1963 no one had actually seen a new island being formed. But in that year fishermen off the coast of Iceland noticed clouds of smoke rising from the sea. At first they thought a ship was on fire. Later they realised that a volcano on the sea-bed was erupting. Before long, steam, smoke and pieces of molten rock were being hurled 4000 metres into the air.

The next day there was a small island where only the sea had been before. The volcano had built itself up above the surface of the sea. For several months, the volcano continued to erupt. Parties of scientists, photographers and newspaper reporters watched the new island growing.

When the eruptions stopped, the island covered an area of 2.6 square kilometres. Its highest point was more than 170 metres above the sea.

Local people named the island Surtsey after the old Icelandic god of fire. Only months after it was formed, the first seeds and young plants were found on Surtsey. They had either been carried there by the wind or been dropped by birds.

Greenland

Umanak harbour in Greenland

Greenland is the largest island in the world. It is part of the kingdom of Denmark. Most of Greenland lies within the Arctic Circle. It consists mainly of mountains and areas of high, flat land or plateaux. The coast of Greenland is rocky, with many steep-sided inlets or fjords.

Snow falls on Greenland in every month of the year. But since the climate is so cold, very little of the snow melts. The snow gets deeper and deeper and turns to ice. As a result, Greenland has the second largest ice sheet in the world, after Antarctica. This ice is always moving. Near the sea the ice is gradually forced through gaps between the mountains. There it forms glaciers. At the coast the glaciers break up into icebergs.

Greenland is one of the homes of the Innuits, or Eskimos. Nowadays, though, many of them have married Europeans. Most of the people live around the coast where the climate is less cold. Their chief work is fishing. Some valuable minerals have been found in Greenland, including uranium and aluminium ore. Mining these minerals is a growing industry in Greenland.

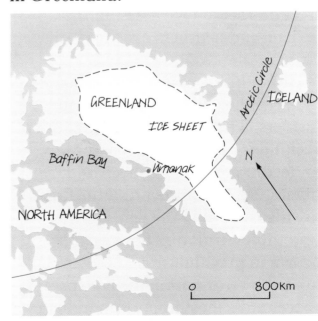

Iceland

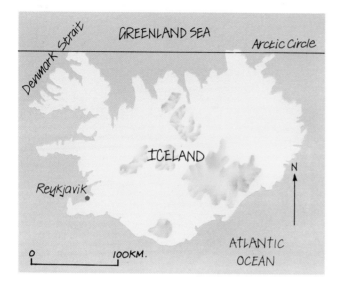

The mountainous interior of Iceland. Can you see evidence of a volcano?

Iceland lies in the North Atlantic Ocean. Most of the island is made up of high, rocky plateaux. Snow-covered mountains and volcanoes, some of which are active, rise from these plateaux. There are also hot springs, and geysers which spurt out steam and boiling water. Occasionally earthquakes occur.

Iceland also has many rivers, lakes and waterfalls. Its coastline has a lot of deep fjords. There are few large trees on Iceland. Grasses, mosses, lichens and heathers are the main plants.

Although there is little fertile soil, many of the people of Iceland are farmers. The chief crop in the south of Iceland is grass. This is cut for hay for feeding the cattle, sheep and horses. Some potatoes and turnips are also grown in the fields. Other vegetables, fruit and flowers are grown in greenhouses. These are heated by water from the hot springs.

Iceland's most important industry

A geyser erupting

is fishing. Some of the fish is frozen or canned for sale abroad. Some is made into fish-meal which is used as a fertilizer.

More and more tourists are now visiting Iceland's beautiful scenery.

28

Newfoundland

Corner Brook Paper Mill, Newfoundland

Much of Newfoundland is covered with thick conifer forests. Felling these trees is one of the main industries. From the trees, timber, wood pulp and paper are made. The saw-mills and factories are powered by electricity from hydroelectric power stations built along the fast-flowing rivers.

Fishing is another important industry. Tiny fishing villages dot the rocky shores of the island. Newfoundland is also rich in minerals. Iron-ore, lead, zinc, copper and asbestos are all mined.

Newfoundland is a continental island off the eastern side of Canada. It is part of that country. The island is situated at the entrance to the huge Gulf of St. Lawrence.

Most of Newfoundland is rocky, with low ground to the east and rugged mountains to the west. The summers on Newfoundland are short, and the winters cold. The narrow strip of water between the island and the mainland is often frozen for 4 months of the year, and icebergs are common. Dense fogs are another hazard to shipping.

Drying fish at Hibb's cove, Porte de Grave Peninsula

Tristan da Cunha

Tristan da Cunha is probably the loneliest island in the world. It lies in the middle of the South Atlantic Ocean. The nearest mainland is the Cape of Good Hope, nearly 3000 kilometres to the east.

Tristan da Cunha is really a volcano. Everyone thought the volcano was extinct. But one day in 1961 it suddenly erupted. To escape, the people had to take to the sea in their boats. After a few days they were rescued and taken to England. Two years later, when the volcano was resting again, the islanders returned to their homes.

Much of the coastline of Tristan da Cunha is surrounded by steep cliffs. There is a small area of flat land in the north-west. It is here that the

village called Edinburgh stands. The islanders fish and grow a few crops such as potatoes and apples. They also keep sheep, cattle and geese. But not many things will grow on such a cold and bleak island. As a result the islanders have to live a simple life.

Newspaper report of the eruption in October 1961

A fisherman off the coast of Tristan da Cunha

The Seychelles

The Seychelles is an archipelago of islands in the Indian Ocean, just south of the Equator. Altogether there are more than 100 islands. Of these, about 40 are made of granite rock from volcanoes. The others are flat, coral islands.

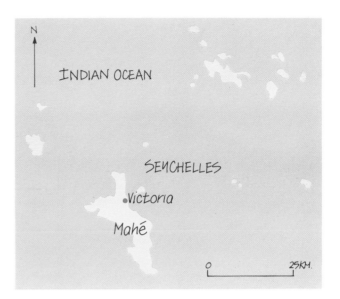

The Seychelles are among the most beautiful islands in the world. Most of them have sandy beaches lined with coconut palms. The main crop of the Seychelles is copra. As we have seen (page 15), this comes from coconut palms. The only town on the Seychelles is the capital Victoria.

A large new airport opened on the largest island, Mahé, in 1971. Since then, many tourists have visited the Seychelles. They go to enjoy the beautiful scenery and warm sunny climate. Interesting plants and animals are found on some of the Seychelles. The coco-de-mer, or double coconut tree, grows nowhere else in the world. The Seychelles are also the home of rare giant tortoises, turtles and frigate birds.

Above: A frigate bird (male)
Above left: A typical beach on the Seychelles
Below: Saying hello to a rare giant tortoise

Mauritius

Rempart Mountain, Mauritius

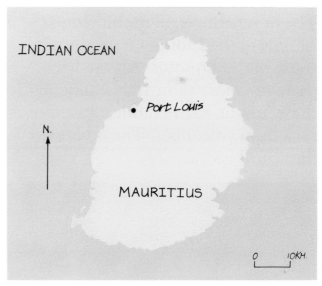

Mauritius is a volcanic island in the Indian Ocean. It is about 900 kilometres east of Madagascar. Mauritius is completely surrounded by coral reefs.

Most of the lower land is taken up with vast plantations of sugar cane. The sugar is the main export from Mauritius. It provides the money to buy in food for the people. Some tea and tobacco are also grown on Mauritius. These crops grow well in the tropical climate. Tourism is another important industry.

Mauritius is one of the most densely populated countries in the world. The poorer people live in very overcrowded conditions. Most of the people are Indian, but some are French, British, African and Chinese. Because of its high population, there is a shortage of farmland on Mauritius. Once the island was thickly covered with forests. But a lot of the forest has been cleared for farmland. As a result, the island's rare wild animals are now in danger. To help save them, large parts of the remaining forests have been made into nature reserves.

Crowded Queen Street, Port Louis, Mauritius

Chamarel Falls, Mauritius

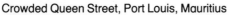

Sri Lanka

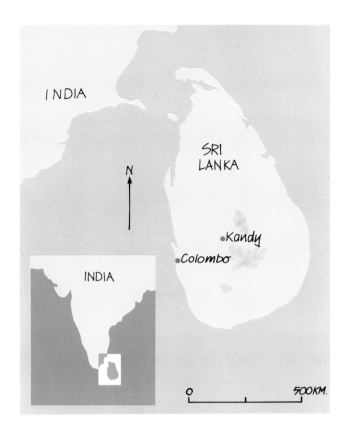

Elephants are trained to move timber

Pickers on a tea plantation in Sri Lanka

Sri Lanka is a continental island which lies just off the south-east tip of India. Much of the coast of Sri Lanka consists of sandy beaches fringed with palm trees. These beaches are becoming increasingly popular with tourists.

Inland, Sri Lanka is mountainous. Since it lies near the Equator, the climate of Sri Lanka is hot. It is much cooler, however, in the mountains. The people of Sri Lanka are a mixture of races. Most of them live and work in the countryside.

In the thick tropical forests of Sri Lanka, valuable trees such as ebony and satinwood grow. These are harvested for their timber. Sometimes specially trained elephants are used to move the logs. Sri Lanka's chief crop is tea. This grows best in the higher, cooler areas. The tea bushes are grown in huge plantations. Each bush is kept pruned to about 1 metre high. Ships take the tea all over the world. Coconuts and rubber are two other important exports from Sri Lanka. Rice is grown in terraced paddy fields on the more gentle slopes of the island. Sri Lanka used to be called Ceylon.

Corsica

A mountain village in Corsica

Corsica is an island in the Mediterranean Sea. It lies in a bay between France and Italy. Corsica has been part of France for nearly 200 years. Its chief towns are Ajaccio, the capital, and Bastia.

Inland, Corsica is very wild and mountainous, but all around the coast there are fertile plains. Sheep and goats graze on the upper parts of the mountains. Since it is such a mountainous island, much of Corsica cannot be farmed. But olives, grapes, oranges and lemons are grown on the coastal plains. Some tobacco and vegetables are also grown. Cork is stripped from the cork oak trees in the forest. The cork is taken from the trunk of each tree every 8 to 10 years. It is exported to factories in other countries. Fishing is another important industry, while Corsica also has some mines and quarries.

More and more tourists now visit Corsica to enjoy the island's warm, dry climate and beautiful scenery.

The cork from these trees has recently been harvested

Corfu

Corfu is a small continental island. It lies just off the coast of Greece in the Ionian Sea. In shape, Corfu is a long and narrow island.

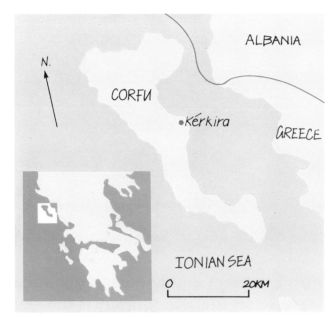

Part of Corfu's beautiful coastline

Going to an island market on horseback

Many of the people of Corfu make their living by farming. They grow olives, oranges, figs, grain and grapes. Goats and bees are the main livestock. Fishing and looking after tourists are two other important ways of earning a living.

Vineyards with scarecrows on Corfu

The interior of Corfu is quite rugged. The hills are almost covered with olive groves and cypress trees. The island's roads are narrow. Many people still use animals to do their carrying and to help with the farm work. Most of the island has not been affected by the large number of tourists who visit Corfu.

In the summer it gets very hot on Corfu. This is why the island is so popular with tourists. In winter cold winds blow from the east bringing heavy rain. This helps to make Corfu the greenest of all the Greek islands.

35

Singapore

Singapore

Singapore is a small island republic. It is just north of the Equator, and the weather is hot and sticky. Much of the coast of the island is fringed with mangrove swamps. Inland there are rolling hills with many small farms.

There are about 2.6 million people on Singapore. Nearly all of them are Chinese. Most of the people live in Singapore City at the southern end of the island. Because Singapore island is so small, no farmland can be wasted. Two or three crops are grown every year on each piece of land. Pigs and chickens are reared in modern factory farms. Even so, much of Singapore's food has to be imported.

One way to get more land is to make it. On Singapore, some of the hills were bulldozed flat. The rocks and soil were spread along the shore line. Factories and houses were built on this new land. Singapore is the largest port in south-east Asia and the second busiest port in the world. It handles much of the trade between eastern and western countries. The cargoes handled include rubber, tin, oil, textiles, timber, rice and many other foods.

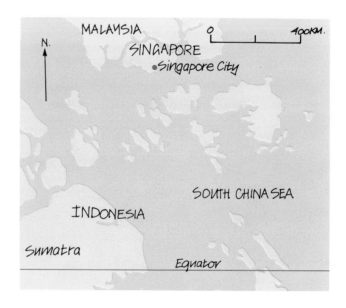

Hong Kong

Hong Kong is not just one stretch of land. It includes Hong Kong Island, the Kowloon peninsula, and a small part of the mainland of China called the New Territories. There are also 235 other islands. The major islands are linked to Hong Kong Island and

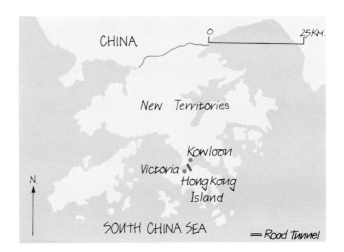

A crowded street scene

the Kowloon peninsula by ferries of all types and sizes.

Hong Kong Island has many tall skyscrapers packed together. It also has a large harbour sheltered by mountains. Two road tunnels and two train tunnels link Hong Kong Island to the mainland.

On most of mainland Hong Kong the people are farmers. Hong Kong also has a large fishing fleet. Even so, much of Hong Kong's food has to be imported. On Hong Kong Island and the Kowloon peninsula there are thousands of factories. Clothes, textiles, machinery, computers, radio and television sets, plastic goods, toys and cameras are just a few of the

things made there. This book was printed in Hong Kong.

Hong Kong is an important centre for trade, particularly with Japan and China. It has the largest container port in the world. But the territory has become very overcrowded. Kowloon has the world's most densely populated housing blocks. Large numbers of people also live on boats and houseboats in the harbours. In addition, the territory has to try to cope with thousands of people from Vietnam and China wanting to settle there.

Crowded houseboats in Causeway Bay, Hong Kong

Trinidad

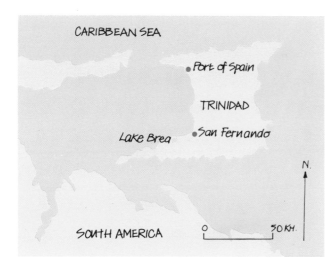

Above: packaging bananas for export
Below left: asphalt worker on Lake Brea
Below right: drilling to find oil on Trinidad

Trinidad is a continental island in the West Indies. It lies in the Caribbean Sea off the coast of South America. Trinidad is about 80 kilometres long and 60 kilometres across. The island was discovered by the famous explorer Christopher Columbus in 1498.

In the south-west of Trinidad is Lake Brea. Unlike most lakes, this one contains not water but asphalt. Asphalt is usually a soft black substance, but when it sets it goes hard. The asphalt is exported all over the world for road making.

Farming provides work for many of the people of Trinidad. The main crops are grown on plantations. They include sugar, cocoa, coconuts, citrus fruits, bananas and coffee. Rum is made from the sugar. The oil industry brings in most of Trinidad's wealth. The oil, like the asphalt, is exported. The tropical climate of Trinidad has made the island a popular holiday resort. Many new hotels fringe its sandy beaches.

New Zealand

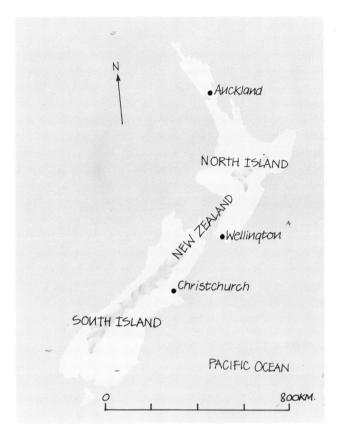

A farm in the hills near Taumaranui

New Zealand lies in the south Pacific Ocean. It is nearly 2000 kilometres from the coast of Australia. There are two large islands, North and South Islands, and several smaller ones.

The two main islands are both mountainous in parts. The central part of North Island is volcanic and has many hot springs and geysers.

The first people on New Zealand were the Maoris. Now these people make up less than one-tenth of the population. Most other people came originally from Britain.

New Zealand's chief industry is farming. About half of the country is pasture on which sheep and cattle graze. Grass grows rapidly in the warm, moist climate. Originally much of New Zealand was covered in

forest. Large areas of the forest were cut down to provide grazing for farm animals. A number of rare and beautiful plants and animals are found in New Zealand. But a great deal of damage to the wildlife has been done by wild and domesticated animals released by human settlers.

Maori children at Marai, near Auckland

Newcomers to islands

Mongoose fighting a snake

Overgrazing by goats on Minorca

When people went to live on an island they often took farm animals and pets with them. Sometimes these animals escaped and became pests.

On the Hawaiian Islands many kinds of wild birds became extinct. This was because of the dogs and cats people took to the islands. Often rats escaped when ships arrived at an island. They quickly became pests. This happened in the Hawaiian Islands and in the West Indies. Later mongooses were taken to these islands to control the rats. But instead of killing the rats, the mongooses killed off all the ground-nesting birds.

Long ago, sailors released goats or rabbits on islands to provide them with supplies of fresh meat. These animals ate all the plants they could find. Many of the islands were turned into desolate wastelands. And all the wild animals which fed on plants had nothing to eat.

Newly planted trees need protecting from deer

Plants can cause problems too. On one Hawaiian island cultivated blackberries have gone wild. They have turned many upland forests into dense jungles. They have made it hard for the island's wild plants to grow.

40

Islands and tourists

Islands are very popular with holiday-makers. Particularly popular are those islands in warm and sunny parts of the world.

On an island there is only a limited amount of land on which food can be grown. Often large new hotels are built for the tourists. The hotels may take up some of this valuable farmland. They can spoil the beautiful scenery the tourists have come to see. The wastes from these hotels may pollute the sea and ruin the fishing industry. Hotels provide work. But often the workers are brought in from other countries.

An overcrowded beach at Las Palmas in the Canary Islands

Often, too, the owners of the hotels live in other countries. This means that most of the money made by the hotels does not benefit the islanders.

The need for new roads, shops and airports for the benefit of the tourists uses up even more farmland. It may mean that much of the island's food has to be imported from elsewhere.

Pollution in American Samoa

This is expensive to do, and not all the islanders are able to afford the food. In addition, other countries do not always have food to spare. In these ways, tourism can sometimes cause problems for island people.

An island airport

41

Do you remember?

1 Which island country is near to the small new island of Surtsey?

2 How was Surtsey formed?

3 What happens to the snow which falls on Greenland?

4 What is the chief work of the people of Greenland?

5 What is the coastline of Iceland like?

6 What is the chief crop grown on Iceland?

7 How are the greenhouses on Iceland heated?

8 To which country does Newfoundland belong?

9 What are the large forests on Newfoundland used for?

10 In which ocean is Tristan da Cunha found?

11 What happened to the people of Tristan da Cunha when the volcano erupted?

12 What are the Seychelles islands made of?

13 What is the main crop of the Seychelles?

14 In which ocean is Mauritius found?

15 What crop takes up much of the lower land on Mauritius?

16 What country is nearest to Sri Lanka?

17 What is the climate of Sri Lanka like?

18 What is Sri Lanka's chief crop?

19 Of which country is Corsica a part?

20 Where is the island of Corfu?

21 Why is Corfu the greenest of the Greek islands?

22 What is the climate of Singapore like?

23 What has been done to make more land on Singapore?

24 What are the buildings on Hong Kong Island like?

25 Why has Hong Kong become very overcrowded?

26 Who discovered the island of Trinidad?

27 What is obtained from Lake Brea on Trinidad?

28 What is New Zealand's chief industry?

29 How may new hotels spoil islands?

30 Why did goats and rabbits become pests on some islands?

Things to do

1 Invent your own island

Invent an island of your own and give it a name. Draw a map of your island. On the map show the capital city and main towns. Draw in a port, and the main roads and railways. Show a mountain range, together with the highest mountain, and three rivers. Use coloured pencils, crayons or felt-tipped pens to help make your map clearer. Give your map a key and show the scale of it.

2 Make a model harbour

Harbours are an important feature of islands. They provide shelter from the wind and waves for ships and fishing boats.

Make a model harbour. Use papier mâché for the cliffs and rocks. Make the harbour walls from strips of wood. Use small cardboard boxes for the buildings, and blue paper for the sea. Paint the tips of the waves white.

3 A model volcanic island

Make a model volcanic island. Perhaps you could make a model of Tristan da Cunha (see page 30).

You will need a piece of cardboard tube about 10 centimetres long and a large piece of cardboard or hardboard. Stick the tube down to the cardboard or hardboard with Sellotape. Mound up clay, plasticine or papier mâché around the tube to make the volcano's cone and the shape of your island.

Paint your model when it is dry. You can make realistic-looking 'lava' by mixing some red or orange paint with a little glue or paste. Slowly and carefully pour this mixture on top of the cone of the volcano so that it runs down the sides. Make some tiny houses to stick down on your volcanic island.

4 Make a working model lighthouse

Lighthouses are built around the rocky coasts of many islands. They help to ensure that ships do not run aground.

You can make a working model lighthouse. You need a clean, empty washing-up liquid bottle. Pull the top off the bottle. Make a small hole in the side of the bottle near the bottom. Thread two pieces of covered wire about 50 centimetres long through the hole you have made, and out of the top of the bottle.

Plug the space around the two wires at the bottom of the bottle with plasticine or clay. Then put a little sand or soil in the bottle, to make it less easy to knock over.

Ask a grown-up to cut off the top of the bottle until a lampholder just fits into the hole. Make a small cut in the top of the bottle on either side of the lampholder. Join up the wires to the lampholder and a battery as shown in the picture. The bulb should light up.

Cover the lampholder with a small clear glass jar. A paste jar is ideal. Paint the tower of your lighthouse. Most lighthouses are white, but some have coloured stripes around them. Lighthouses also have a door and small windows in them.

Your lighthouse would be easier to use if you had a switch to turn the light on and off. Can you make a simple switch like the one in the picture?

5 An island expedition

Imagine that you are the leader of an expedition which is going to explore a remote island in the South Pacific. Write a story about your expedition, mentioning what preparations you make for the trip, what equipment you take, how you reach the island, and what you find there.

6 Find some European islands

Trace an outline map of Europe into your book. Mark on it where you live. Now put on the map some of the islands you know of. Which of these islands is nearest to your home? Which is the most distant? Which is the largest of these European islands?

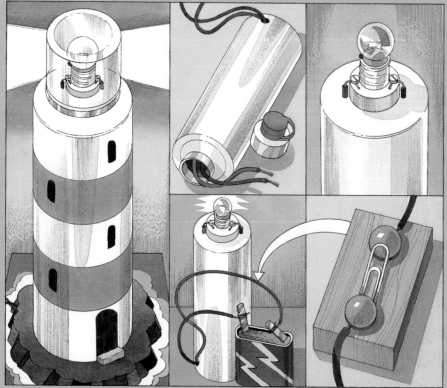

7 Islands and tunnels The island of Anglesey is linked to the mainland of Wales by a bridge. Tunnels link Hong Kong Island with the mainland. After very many years talk, a tunnel is being built to link Great Britain and France.

How strong are different tunnels?

You will need some thin card, scissors, Sellotape, 3 yoghurt pots, drawing pins, a large piece of wood, glue, and some coins or beads (all the same size).

Cut three pieces of card 30 centimetres long and 12 centimetres wide. Make three different shaped tunnels with the card. Use only four drawing pins in each tunnel.

Carefully glue a yoghurt pot in the middle of the top of each tunnel.

Put coins (or beads) in one of the yoghurt pots. How many coins can you put in the pot before the tunnel collapses? Do the same thing to the other two tunnels.

Which shape tunnel will support the most weight? Is that the shape you most often see being used in real tunnels?

8 Pictures of fishing boats Collect pictures of the different kinds of boats and ships which are used for sea fishing. Stick your pictures in a book or make a wallchart of them. Write a sentence or two about each one.

9 Holiday islands Choose one of the islands people visit for their holidays. Collect pictures of that island. Look at holiday brochures, magazines and travel agents' advertisements for help.

Arrange your pictures in two groups. In one group put pictures of the island which would make you want to visit it. In the other group, put pictures of features of the island which you find unattractive.

Make a book or wallchart with your pictures. Write a sentence or two about each one saying why you like or dislike it.

10 The bottle on the beach Some people used to throw bottles with messages in them into the sea. They hoped that one day a reply would come from a distant land.

Pretend that one day you are walking on a beach when you find a bottle with a message in it. The message says, 'Help, I am stranded on Craigmouth Island'. Write a story describing what you do and what happens. . . .

Things to find out

1 Here is a map quiz. An atlas will help you to answer these questions about the map below.
(a) The map shows a group of islands. What is the name of the country they form?
(b) What is the name of the city marked on the largest island?
(c) Is the largest island flat or mountainous?
(d) There is one large peninsula. What two countries are situated on this peninsula?
(e) What is the name of the sea between the islands and the mainland?
(f) What is the name of the ocean to the south and west of the islands?

2 On Hong Kong and some other islands with large populations there is a shortage of drinking water. This is in spite of the fact that islands are surrounded by water. Is it possible to obtain drinking water from sea water? If so, why is it not done?

3 Mention was made on page 14 of the British scientist Charles Darwin. Find out more about the life and work of Darwin.

4 In what ways is a tunnel better than a bridge for linking a continental island to the mainland? In what ways is a bridge better? How would building a tunnel or bridge linking a continental island to the mainland, make life easier on the island? How might the bridge or tunnel make life on the island more difficult?

5 Rice is an important food crop which is grown on several tropical islands. Find out where rice is grown and how it is looked after. How does the mud carried by rivers help rice to grow?

Terraced rice paddies in Sri Lanka

6 The dodo was a large flightless bird that lived only on the island of Mauritius. The last dodo seems to have died in 1681. Find out all you can about the dodo and why it became extinct.

Glossary

Here are the meanings of some words which you might have met for the first time in this book.

Archipelago: a group of islands.

Asphalt: a black, tarry substance used in road making.

Atoll: a horseshoe-shaped or circular reef of coral around a lagoon.

Continent: one of the seven large pieces of land on the Earth's surface.

Copra: the dried flesh of the coconut.

Coral: chalky skeletons of tiny sea animals called polyps.

Drifter: a boat which catches fish that live near the surface of the sea using a net which hangs down from floats on the surface.

Equator: the imaginary circle around the centre of the Earth.

Evolution: the slow, gradual change of plant and animal species.

Fjord: a steep-sided valley carved by a glacier and flooded by the sea.

Geyser: a spring which sends out hot water and steam.

Glacier: a large river of ice which flows down a valley.

Hydroelectric power station: a power station which uses the energy of running water to help make electricity.

Ice Ages: the periods of time, thousands of years ago, when the Earth was much colder than it is today.

Iceberg: a large block of ice which floats in the sea around the polar regions. The ice was originally formed on land.

Ice sheet: a huge sheet of ice and snow covering Greenland or Antarctica. During the Ice Ages an ice sheet covered large parts of Europe and North America.

Island: a piece of land with water all around it. Oceanic islands lie far out to sea. Continental islands always lie close to the mainland.

Lagoon: sea water enclosed by a coral reef (also a shallow lake separated from the sea by a sandbank).

Lava: the hot, liquid rock which comes out of a volcano.

Ore: a rock that contains a useful mineral.

Paddy field: a flooded field in which rice is grown.

Plantation: a large farm or estate which specialises in crops to be sold, such as rubber, sugar cane, cocoa etc.

Plateau: an area of high, flat land.

Polyp: small jelly-like sea animals whose hard chalky skeletons form coral reefs and islands.

Purse-seining: a method of fishing in which a shoal of fish is surrounded by a net which is then closed underneath them.

Reef: a bank of rock, sand or coral off a sea-coast.

Stack: an isolated mass of rock with steep sides which rises from the surrounding sea.

Tourists: people who travel from place to place during their holidays.

Trawler: a ship which catches fish (that live on or near the sea-bed) by towing along a large bag-like net.

Volcano: a weak part of the Earth's crust through which molten rock or lava from inside the Earth comes out.

Index